WHAT IS IT? CREATION OR EVOLUTION

The answer to man's foremost question

Thato Lerato Mohlathe

ISBN: 9798852156679

CONTENTS

DEDICATION
TO MY CHILDREN

To my son Katlego Mohlathe and daughter, Bontle Mohlathe, mom loves you a lot. I dedicate this book to you that you may learn to nature your talents and have the courage to take risks.

INTRODUCTION

Human beings are naturally intelligent creatures varying by talents and capable of reasoning and logical judgments. Though so, it is often left to the scientific pundits to tell the story of the origin of our planet and its living things. Whilst our world is made up of a provider with expertise in his or her field and a consumer, it is reckless to give limitless ownership to the scientific community to sell its idea of the origin of our planet because after all, they are human.

A biased position is inevitable when engaging the title of this book. You can only represent one of the two concepts. With that said, the champion will be the one who presents facts and is able to adequately substantiate their concept of choice. And the audience should be able to differentiate facts from opinions to make sound judgements.

The most critical factor of a theory lies in the lens of its concept. A lens can lead to a fragmented outlook when it disregards other realities that may exist within its ambit. So far as the foremost question in human existence is concerned, the glorious universe and its power remain subject to mere human intelligence and imagination. It can neither be well explained by scientists nor

Christians without leaving a question mark. How close are we to answering this question? Well, we should rather look at which answer is the best.

CHAPTER ONE
NO END TO THE DEBATE

There's an ancient debate and claims of scientifically proven investigations about the origin of man. The most recent and prominent is the argument between the leaders of the Intelligent Design and Evolution theory with no end to this discourse. Scientific theories testify of a beginning, millions or even billions of years ago which no human has seen. Religious communities see these theories as "personal opinions" and flirtation of the imagination by scientists who exploit the science rubric to advance their beliefs or invalidate the existence of God. They believe some scientists aim to manipulate members of the general public who have faith in its credibility. On the other hand, we have Scientific Creationism that aims to prove that God created our planet out of nothing.

According to the Intelligent Design theory, the complex life forms found in the universe cannot be explained solely by natural causes, hence an intelligent transcended entity contributed to the origin of the universe (David L. Hudson Jr.).

On the other hand, the evolution theory postulates various types of plants, animals and other living things on Earth have their origin in pre-existing types and that the distinguishable differences are due to modifications in successive generation. (Francisco Jose Ayala).

I hereby employ the *creative thinking* outlook for the examination of our environment, simply put; to investigate elements of creativity in our planet to substantiate the creation theory notwithstanding that nature has complex and unexplainable features such as reproduction that cannot be compared to the ability of man and creativity of manmade inventions. With that said, there are standard comparable basics that can be examined through the lens of creation and creativity.

We can all attest to the fact that creativity is one of our planet's foremost characteristics found in nature and man-made things. Today, unlike 500yrs ago and beyond, man has created his own inventions namely; cars, planes, cell phones, laptops, stoves, refrigerators, televisions and many others. Human beings have proven to be certified creators of sophisticated inventions with complex systems.

Man has acquired knowledge and an invention formula out of which he can develop a blueprint to study whether his environment has the attributes of creation or not. We are now competent to participate in this exercise or even develop more efficient methods of investigation through our experience and knowledge as creators able to identify the features of things which are created.

We are inheritors of the heroic work of Stove Jobs for the iPhone, Thomas Edison for the electric light bulb, Henry Ford the developer of the assembly line and the Wright Brothers who invented the airplane. All bringing forth the greatest civilization in the history of man that has changed our social, economic, skill and knowledge life.

Let's characterize an invention to see if earth fits into this description. An invention is a whole complete in itself and produced for its function. Sophisticated inventions have complex systems which are interconnected to complete the whole. They are the building blocks of an invention. For example, a car has an exhaust system, fuel system and others. Is the body an invention because it too has systems such as the digestive system, immune system and others?

The invention blueprint is found in all inventions though it may vary based on the type. Cars have building blocks, we call them "car parts", the body also has them, and we call them "body parts". The inventor must have absolute knowledge of the invention and the environment it will operate in. For instance, a car's headlights exist because of darkness in the environment it has to function in.

In the case of our bodies, we have eyes to see our environment. The environment is one of the key determinants of the inventions design. Some designs on the invention are determined by the environment it has to function in. For instance, windscreen wipers are part of a car's design because of rain in the environment. And ears are part of the body because of sound in the environment that may come from a television or someone talking, so on and so forth.

An invention with a design determined by the environment it will function in displays that knowledge was acquired and used, for example, the knowledge of sound in the environment for an ear or rain for a windscreen wiper. Planet earth proves to be designed by an entity that used knowledge. When we compare nature's with man-made design we discover similarities from systems, to designs determined by the environment and building blocks. This implies that there is a standard formula in things that are created under the sun.

CHAPTER TWO

ARE ANIMALS AND PLANTS FOOD OR
ARE THEY RAW MATERIAL

The United Nations reports an increase in the global population from an estimated 2.5 billion in 1950 to 8 billion in mid-November 2022. This means 5.5 billion bodies were reproduced1950-2022. https://www.un.org/en/global-issues/population#:~:text=Our%20growing %20population,and%202%20billion%20since%201998.
Animals and plants play a critical role in producing and replenishing our bodies though they also provide pleasure. Likewise, inventions are created for their function; however, pleasure and beauty are often an add-on. Let's take a car for example; it was invented for transportation, to move the passenger from point a to b. Cars come with extra features for pleasure and beauty, like the music system and the colour of choice. These do not affect the function of the car, whether it moves from point a-b, they are an add-on.

The add-on formula is in all inventions or work of tangible creativity. First is the structure and related function, and then second is the pleasure and beauty/decoration. The invention through its structure and related function can produce its intended function without an add-on. Nature also has the add-on formula, like pleasure and beauty. Our food is one of them, its function is to produce and replenish our bodies but simultaneously provide pleasure through flavour detected by our taste buds, like the chocolate many of you love. Eating is compulsory for survival which is the sole purpose but it would be frustrating to perpetually eat foods without the add-on feature, for this reason pleasure was included to the experience.

As we focus on creation by comparing the design and creativity of man-made tangibles to establish a formula and survey our environment to find whether it fits into this description, we look at the production of a whole which is complete in itself, the body, which we reproduce as so then the matter that produces it is raw material. Are animals and plants just food or are they raw material? We hereby use a technical lens to study functions though there are other features to food (animals and plants).

The purpose of raw material is to build a product and that's what animals and plants do, they build the body though we have not termed it a product, an etymology matter. The word "food" does not reveal a function hence this takes place in our subconscious mind also owing to the fact that we get lost to the add-on feature of pleasure and beauty.
It is in the stage of pregnancy where nutrients show themselves to be raw material through their role in the creation of a new body.

The body and the environment may look like two independent entities yet they are connected, this is captured in the ecosystem concept. The body is a whole complete in itself by design yet it is not independent. For a biological outlook, the body can travel back and forth planet earth, this makes it look like an independent entity but the globalization of animals and plants-living and non-living things makes it possible. Rice found in Africa is the same rice in America, keeping the body within the system which it is connected to for life. Consequently, the idea that man can live on other planets in the universe is a myth.

CHAPTER THREE
THE SPIRIT MAN'S ROLE IN BIOLOGY

The danger of the traffic of ideas, knowledge is not always facts it can be false. Let's work on Charles Darwin attempt to explain why children look different from their parents in a concept called "variation". Our analysis reveals that planet earth embodies an entity that used knowledge. Also, we have studied that the inventor has to have absolute knowledge of his invention and the environment it will function in.

There is one factor that Charles Darwin did not consider which is the spirit is being and his traits. Not every body organ exists for a direct biological function. The direct biological function of the heart is to pump blood. With that said there are biological organs that carry out *direct* functions for the speaking and hearing spirit though the whole body is for the purpose of the spirit's life on earth. The function of the ear is for the hearing spirit, and the voice box is primarily for the speaking spirit. In that regard, the variation concept's outlook is incorrect.

There are biological aspects that are intended to exist for the direct function of the spirit being.

What would it look like if 8 billion human beings on our planet looked like Donald Trump? How can the entity know that there are sounds to be heard, objects to be seen, and oxygen to be breathed, and not know that the spirit man is an individual?

Whose individuality is based on the fact that there will be a population? Charles Darwin's view is deficient in knowledge; it does not consider the spirit man as a factor in determining the biology of the body and is too narrow a lens whereas the entity's knowledge must be absolute, vast and encompass population and generations to come.

The knowledge the inventor must possess is that the spirit man is an individual to dwell among others and they must be able to distinguish each other. Just by The knowledge the inventor must possess is that the spirit man is an individual to dwell among others and they must be able to distinguish each other. Just by understanding that the planet is majestic and powerful by design we must also know that it wouldn't be difficult for an entity that created the universe to create humans that will look different as an intended design.

CHAPTER FOUR

THE REPRODUCTION SYSTEM AND THE FEATURES IT SUSTAINS

Reproduction is a phenomenon that cannot be found in man-made inventions or tangible substances that are created. It remains to amaze man; it looks so supernatural and unexplainable. In this book we will learn about the different features that the reproduction system enables as its purpose which without cannot materialize. One of the features that the reproduction system makes possible is the continuity of man's existence for thousands of years. Once we identify that there is a system in place whose efficacy is proven by the function it produces, then concepts such as natural selection don't hold water.

Continuity is an ideal word to best describe the role of the reproduction system in the making of generations of the human race. The reproduction of plants and animals is an on-going supply of raw material for our bodies.

Due to the reproduction system, the continuity of mankind is guaranteed. The discontinuation of reproduction is the only logical element that can cause the extinction of man. If plants and animals stop reproducing today, then there will be no supply of raw material to reproduce the bodies in the womb and sustain them after birth. Therefore, the reproduction system is central to the continuation of man for generation.

Another feature that the reproduction system enables is the preservation of man, plants and animals. The secondary to producing living things and man in particular, the sole purpose of the reproduction system is preservation. Raw material is preserved in the reproduction system. The creation or the start point of that creature is what defines what type of genome that will be preserved by the reproduction system.

Merely having the genome as the code does not guarantee that it will be passed on to other generations. The reproduction system is central to the processes that are encompassed in the preservation of the creature. What the Evolution theory lacks is the breakdown of the bricklaying mechanism that forms man, animals and plants. Varying from your mother will never stop you from being a human being biologically. This aspect of our biological being remains intact and those that evolve were designed to, there is evidence for such from human beings that existed B.C.

Eternity is another feature the reproduction system proclaims. We can all attest that there is an on-going supply of raw material (animals and plants) as a result of the reproduction system that is why man has lived for thousands of years. Human beings come and go but the earth remains the same and is faithful in its supply. This demonstrates that from the two, the environment and man, the problem lies with the biological aspect of man because the environment will go on to produce and sustain new bodies.
To recap, these are the three important features the reproduction system makes possible

- On-going provision of raw material for new bodies and to sustain life
- Perseveration of man, animals and plants from their ancestors
- Eternity

CHAPTER FIVE
PRIMARY AND SECONDARY INVENTION

Earth proclaims its traits of creation. Anything that is created will have its set of laws such as nature's. What is earth's purpose? Can earth reveal this secrete either than simply getting it from the scripture? The broader view of the reproduction system entails raw material (animals and plants) to work together with sperm and ovary in the reproduction of man reveals earth's end-product. This end-product reveals the purpose which is the reproduction of man, being the supreme creature. The add-ons such as pleasure and beauty create the experience. When we observe life from its final point after processes of the reproduction system producing it and the add-ons, a human experience is its goal.

The human being through his spiritual traits has invented material that has added lifestyle value to his human life experience.

Earth's primary purpose is to bring the man into existence and provide supply of raw material from womb to tomb to replenish his body whilst he through solving his problems he has invented lifestyle material.

The "Primary and Secondary Invention" is a concept that describes earth as the primary invention and material created by man as the secondary invention. The primary invention is for the purpose of life, it is made out of living and non-living things such as plants, animals, water, oxygen and others. The secondary invention consists of material lifestyle products invented by man such as cars, laptops, cell phones and others.

Man-made inventions attest to be a product of the spirit man. The spirit is the inventor that produces ideas, methods, strategies, creativity, thought and talent. It's irrefutable that the spirit man is a creator. This experience displays the type of entity that would be required to create planet earth.

CHAPTER SIX

SYSTEM

Man-made systems demonstrate that a system can only be a substance created, it cannot just happen. The complex processes of a system are attributed to thought, intelligence, and skill of the spirit man.

Systems are a set of things interconnected to complete the whole. The building blocks are selected according to their function. This implies that there must be a selection process. Additionally, it must produce a whole. The efficacy of the whole exhibits an intended design.

If one building block is removed from the system the whole will cease to function. For example, if an alternator was disconnected, a car will stop functioning. This also applies to earth, if water ceased to exist, the extinction of man and other living things would be instant.

Earth is a complete whole for the purpose of life. It has different systems such as the ecosystem, the body's systems, the reproduction system in plants and animals and more. Through the experience that we have with our man-made systems, we can attest that systems are a phenomenon that can only be created.

CHAPTER SEVEN

DESIGN SPECIFICATIONS

To illuminate the fundamentals that reveal earth as a creation we will explore the scientific community's failure to cure cancer because it circumvents the laws of nature. The term "foreign body" that describes synthetic drugs can be an ambiguous concept when after all; all ingredients originate from our surrounding environment.
What are the characteristics that make a drug a synthetic drug?

The Creation Theory framework is the perfect outlook to address the ambiguity found in the "synthetic drug" concept because all inventions—the whole that is complete in itself have specifications. Raw material specifications are the ingredient—structure with which an invention—the whole is built. Nothing can become its specific invention by means of generalization in building blocks that build it.

A "particular specification" is a phenomenon that does not exist in unit form external to an invention but becomes one should it be used in building an invention. In an invention, specifications have the phenomenon of "particular" because they form the building blocks that are interconnected in complex network to complete the whole.

The environment consists of a cow, apple, water which

have specifications—structure and related function. Taking into account that the body is reproduced from its environment it means it has been built by its environment specifications. It has its environment specification from which it was reproduced and is sustained by it. *The harmony between the body and the environment is simply because the body is produced from it.* Like water and the body or the cow that eats the grass and the body.

A synthetic drug is a structure that the body is not produced from. Inventions have specifications and a synthetic drug is a specification not found in the body and if not found in the body it is therefore not found in the environment in its modified or reinvented form.

CHAPTER EIGHT

SYSTEM OF ABUNDANCE

This segment focuses on abundance which passively takes place in the background, lacks awareness but is now revealed. When we study the determinant of abundance in nature we do so by examining the function of the source—the seed and not food in the supermarket. We examine the seed; and study its invention specifications and function. We can in this case use the apple seed as an example to investigate whether food shortage in the context of population growth corresponds with the system of existence. When one seed produces a tree that produces 50 apples that have seeds that can germinate and produce 35-40 apple trees that will produce more or less 1700 apples that have seeds what do we call this system?

The fact that one seed alone has the capacity to produce a tree that can produce fifty apples is by construct an expression of a system of abundance.

This observation is not about which seed will germinate or how many will of the many or whether man exploit this feature of abundance to maximum capacity or not but rather to activate the knowledge of its existence and not to be ignorant and deceived. It also serves to elevate this reality into the perceptions of man and develop a framework to have a sound thought amid a plethora of concepts that contradict the reality of existence among others; the suggestion that man must colonize other planets for survival

(Stephen Hawkings), this suggestion although the fish is still reproducing as it has for thousands of years. Central to the guarantee of eternal supply of raw material is the reproduction and abundance system which is embedded in the design.

This system of abundance is paralleled with our reproduction, continuity and most importantly population growth. Due to demand, this generation farms more plants and animals than previous generations to meet the demand of the population growth.

If we are to measure the significance of the invention specification —stored in the genome of plants we recognize that if we were to eliminate the abundance feature from the seeds there would be the real kind of food shortage and probable extinction. We are currently sustained by this feature which may be expressed differently across plants but is an existing phenomenon.

CHAPTER NINE

ADAM AND EVE VERSUS EVOLUTION

The Intelligent Design and Evolution theory's debate is mainly centred on the origin of man. The most feasible method of solving this debate is investigating the male and female phenomenon. If there can be no evidence as it is, truly speaking to envisage the beginning of man to have come by through the creation of Adam and Eve is more sensible than the idea that man evolved from an ape. The ape story does not correspond with the fact that man is made up of the following unchanging feature of creation;

- Reproduction
- Building blocks
- Systems
- Raw material
- The spirit man and his attributes
- Design Specification
- Designs on the body determined by the environment (eye, nose)
- Designs on the body directly determined by the spirit being (ears)

Though Christians will always be challenged to prove that there is a deity who created the universe, the idea that he is superior is part of critical thinking to equate the admirable and powerful planet to an ability not found on earth.

For example, it would be absurd and stupid to say a dog invented a car, never mind that the human being is nowhere to be seen but it remains that it is impossible for a dog to invent a car. This way we eliminate the impossibilities to find the possibilities. It is important to understand the logical thinking of Christians, we do not attribute our planet only to the biblical scriptures but by observation, it requires a transcended entity to have formulated it.

For this reason, the Evolution Theory remains a myth. As first nature, we cannot escape our instinctive observation of our planet as creators ourselves. We can see the formation of creativity in our environment which we are so familiar with. Our spirit man is programmed to scanning his environment according to his ability and traits of his experience such as acquiring knowledge over time.

CHAPTER 10

KNOWLEDGE

As we endeavour to use a model created out of our experience as creators to explore our environment we should capture the essentials that are important in an invention process. We can make an evaluation in nature about one of God's characteristics as being omniscient and omnipresent. Our experience as creators reveals that knowledge is vital in the creation process to determine the inventions design and specification. We do not only need knowledge of the resources to be used but the environment as well.

In a car for instance, knowledge of whom it created for in terms of the driver's body structure to have things such as brake pads and why the need for such, the knowledge of where the car will operate, are all part of the absolute knowledge that is required.

When a theory is presented through a lens without other existing realities that may debunk it, it sounds real. It is evident that the planets creator must possess absolute knowledge and think in concert with other factors in mind to design a sustainable and efficacious material. At the top of our mind we see an entity with a brainpower that covers the globe, even the universe and knowledge of all its parts. We can confirm this through breaking down the components that build nature, they can tell us what was known.

The general view is that Evolution is a theory that was conceived to refute creationism. Therefore, can creation defend itself against this attack? For sure, if it is to point out the structural elements of creation and shift from a faith based approach of the resurrection of Christ.

.

However there are two key weaknesses of Charles Darwin's theory and one is his small imagination and the other is the primitive time he lived in just before the advent of many of our inventions. His aim to refute creationism focused his lens on a viewpoint that cannot be substantiated though some part of the scientific community has endeavoured to substantiate the theory through Molecular Biology. This viewpoint left out some elements that can only be explained by a creation theory.

CHAPTER 11
CREATIVITY CULTURE A FEATURE OF CREATION

Is something creative something created? Earth is creation colonized by creators. The fact that earth is colonized by creators can be proven by compiling a long list of man's inventions or intangible items that undergo the process of creating. Our planet provides natural resources as raw material for the creation of man-made material. Nature shares a similar formula for substances that are created such as building blocks, systems, design specifications and more.

Let's use an analogy of the ocean to depict our experience of earth as creation and creativity, take for example the ocean, if man lived there their activities would be accustomed to their environment. What they could then do as man in the ocean would be limited to the resources it avails and their problem solving skills out of which many invented substance originate from. Similar to this experience on land, man's creation and creativity are limited to what nature provides coupled with their problem solving skills.

As we have learned in one of our chapters that the spirit man plays a significant role in his biological existence, one of the important aspects of man's existence is the things which he has received but did not ask for, starting with the gift of life itself. To break it down, he neither asked for his biology, his spiritual formulation nor had any role in choosing it such as his gender or race or whether he would be a hearing or speaking spirit or not.

This reveals that there's a fixed nature in place among others such as an inherent feature for the spirit man is to recognize beauty or pleasure. This is harmonized between the environment and the spirit man such as raw material (animals and plants) is for his biology so is his love for colour and all things beautiful.

The language of creation and creativity is in the tongue of man and the interpretation of poverty is not the monetary value but the consequence of lack of affordability for material assets, delicious foods and variety thereof. Here in my country the indigent live in shanties, man look and say they are poor simply because they do not live in houses with beautiful tiles, a bathroom, finest wall paint and exquisite furniture with an exceptional design. These are the characteristics of living under the sun which are fixed and man is bound to, the creativity culture that is inherent and dominates our world. However the reality takes place in the subconscious mind of man its meaning is not clearly seen because it is not well explained.

But you can't trust the man's judgment of colour, the pink one from Europe calls himself white and the brown one from Africa calls himself black.

CHAPTER 12
CONCLUSION

Nature indisputably possesses similarities of creation, creativity and design as man-made inventions. From building blocks, to systems, to design specifications, designs determined by the environment, designs determined by the speaking spirit, the reproduction system, preservation, eternity, add-ons, the abundance system and the use of absolute knowledge.

The above elements that have been outlined as part of nature's formulation do indeed prove that man, animals and plants couldn't have come into existence through an uncontrolled process such as natural selection.

The efficacy of lives and role of living and none living things demonstrates purpose. You don't need to be told or written a manual about what a car was created to do. All you need to do is experiment with it and drive it; you will discover what it was meant to do. The efficacy and function will reveal its purpose.

Therefore, it could be that the creator left us to explore and discover our planet knowing that we will have the knowledge and intellectual capacity to do so.

ABOUT THE AUTHOR

Thato Lerato Mohlathe

Thato Lerato Mohlathe originates from South Africa. She was born in exile, Swaziland, to Ben Gerald Mohlathe and her late mother Beauty Maria Motsepe in 1983 to join her three siblings. Shortly after, the family moved to Tanzania where she spent her early childhood years. Her family was among the many that repatriated to South Africa after the release of Nelson Mandela and the fall of the Berlin Wall. Thato's intellectual attribute and school of thought was influenced by her political experience. A dropout of African studies, politics played a significant role in shaping Thato's viewpoint of life and interests but it wasn't until she participated in the American "discovered" HIV/AIDS research that she was introduced to the scientific community, its practices and concepts.   She discovered an outlet for her innate capacity as a critical thinker and combined it with her writing talent after decades of struggling with conformity in the African continent plagued by ignorance and somewhat intellectual bankruptcy.